I0465182

# CLOSED, YET IN MY RIGHT MIND

## 2nd Edition

Dr. Verdree B. Stanley

authorHOUSE®

AuthorHouse™
1663 Liberty Drive
Bloomington, IN 47403
www.authorhouse.com
Phone: 1 (800) 839-8640

© 2017 Dr. Verdree B. Stanley. All rights reserved.

No part of this book may be reproduced, stored in a retrieval system, or
transmitted by any means without the written permission of the author.

Published by AuthorHouse  05/22/2017

ISBN: 978-1-5246-9310-7 (sc)
ISBN: 978-1-5246-9309-1 (e)

Library of Congress Control Number: 2017907910

Print information available on the last page.

Any people depicted in stock imagery provided by Thinkstock are models,
and such images are being used for illustrative purposes only.
Certain stock imagery © Thinkstock.

This book is printed on acid-free paper.

Because of the dynamic nature of the Internet, any web addresses or
links contained in this book may have changed since publication and
may no longer be valid. The views expressed in this work are solely
those of the author and do not necessarily reflect the views of the
publisher, and the publisher hereby disclaims any responsibility for them.

King James Version (KJV)
Public Domain

# Contents

# Preface

It has been a great concern of mine to explore areas of secular and biblical perspectives of mental illness. As humans our thinking, mood and behaviors are all continuing until death. Being that mental illness covers a wide range of health conditions. One may ask, "Where does this illness/disease come from? Is it inherited? Will each individual have different experiences with the same diagnosis? According to NAMI (National Alliance on Mental Illness), one in four adults in America experience mental illness in a given year. Approximately 20 percent of youth ages 13 to 18 experience severe mental disorders in a given year; among children ages 8 to 15, it is estimated that 13 percent were diagnosed as having symptoms of mental illness.

If being, closed, yet in my right mind; blocks me from my goals of entry, then why exist? Is it possible to live and emerge from my present state? The path was dark but only up until the present. I ask myself, where does God fit into the mold? Or can he fit? Where is God in all of this? In this 2nd edition of my book I want to see and explore God's word as it relates to mental illness and the wholeness of humans.

# New features in this 2nd Edition

1. An analytical view of mental illness.

2. A biblical view of mental illness.

3. At the end of each chapter there is a secular crossword puzzle and a biblical crossword puzzle pertaining to information in the given chapter.

4. At the end of the book are the answers to the puzzle.

5. When completing the secular crossword puzzle you may want to refer to the article.

6. When completing the biblical crossword puzzle you may want to refer to the King James Bible.

7. All of the articles in this 2nd edition can be located on Google; (Google it).

# Chapter One

## Standing on your own two feet

In the article entitled "The History of Mental Illness: From Skull Drills to Happy Pills", Allison M. Foerschner, gives a historical progression of mental illness, it's treatments, reconstructed event, customs and styles from the past to present day results. These developmental milestones are a set of functional skills that enables a person to function in a normal capacity. Every intended goal is a way of achieving an aim or solving a problem by trying different methods and learning from your mistakes.

1. Treatment: The techniques or actions customarily applied in a specified situation.

2. Equilibrium: A state in which the mind, body and soul, opposing forces are balanced.

3. Caged: A physical and/or mental structure to keep individuals captive.

4. Bad Fate: Something that causes or entails suffering.

5. Asylums: An institution offering shelter and support to people who are mentally ill.

Sometimes, standing on your own two feet can be a challenge. Is standing alone a result of a person not being healthy? What happens when the human mind turns against itself? And what can be done to solve these problems.

Every individual grows and develop at different rates no matter the age of the individual.

1. Social skills;

   Being able to convey one's thoughts and ideas may be the single most important skill that you can develop in life.

2. Emotional development;

   Love and support are positive growth in your achievements with support of others. Give yourself opportunities for self evaluation.

3. Cyber bullying;

   The internet, cell phones or other devices used to intentionally harm others through hostile behavior.

4. Being aware of your limits;

   A rule or circumstance and the quality and state of being.

# The Establishment

Before there was action in a given space, a starting point in a given moment or giving attention to an ideal, God "was" and still "is" (Genesis Ch., 1: verses 1-5).
God did not become God; he is God. (John Ch., 4: verse 24).

His omnipotent power allow him to be everywhere at the same time. (Psalm 139: verse 8).

God is the beginning; he put time into motion. (John Ch., 1: verses 1-4).

God is the end; he will command time to end. God is the creator of life and death. (Revelation Ch., 22: verse 13).

The ability to confer and invoke divine favor for family, relatives and others; asking God to care for and protect someone or something is a necessity. This process can be achieved through, 1. Giving God praise, 2. Worshipping God, 3. Glorifying God, 4.Giving God all honor, 5. Exalt God. Using these five maneuvers coupled with belief and faith in God will manifest positive outcomes.

# Secular Crossword Puzzle
## Exercise - Chapter One

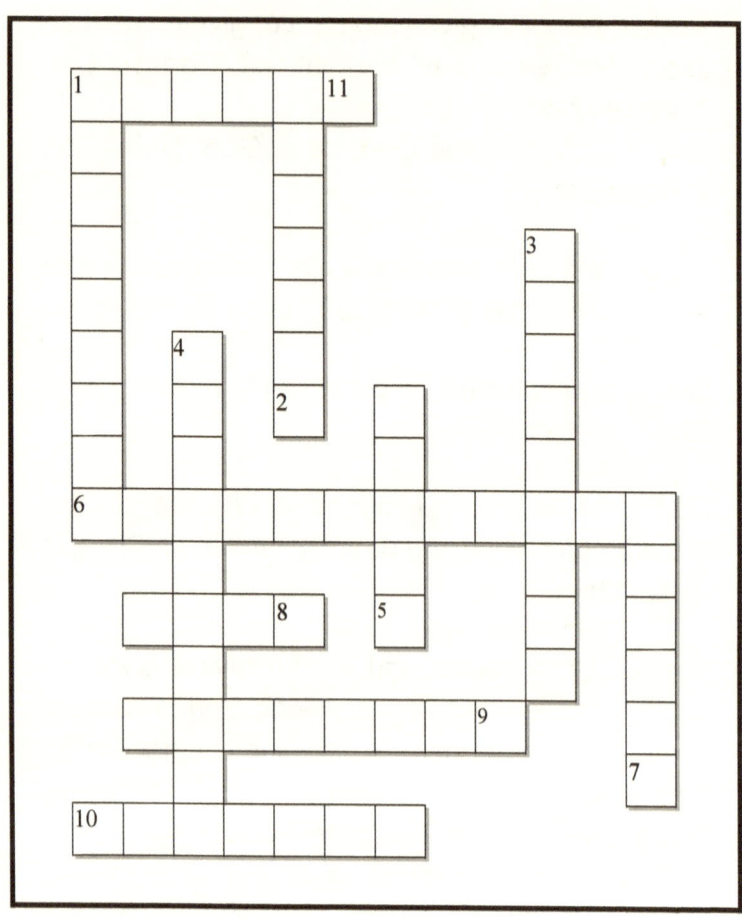

## Across:

1 - _____; An institution for the maintenance and care of the mentally ill.

6 - _____; Force, displeased, deity, mental illness.

8 - An_____eye or an angry deity.

9 - _____; Conversation disorder.

10 - _____; The first mental hospital was established in 792CE.

## Down:

11 - _____; were established and run by members of the clergy to treat the mentally afflicted.

2 - Human_____; Threats, bribery, punishment, submission.

3 - _____; Lengths, metric, imperial, units.

4 - _____; also referred to as, trepanning.

5 - Blood, phlegm, bile, _____bile.

7 - _____Illness; It dates back as early as 5000 BEC.

# Biblical Crossword Puzzle
## Exercise - Chapter One

## Across:

2 - I am alpha and_____,

3 - In the_____God created the heaven and the earth.

5 - _____things were made by him;

6 - And God_____,

8 - And the word_____ with God,

10 - The_____and the last.

12 - Thou art_____;

## Down:

1 - And God_____the light from the darkness.

4 - And darkness was upon the_____of the earth.

13 - And the word_____God.

7 - And the life was the_____of men.

9 - God is a_____,

11 - If I make my bed in_____,

# Chapter Two

## Who says I'm wrong?
## Who says I'm right?

Being able to direct a person or behavior towards a path of independence you become more powerful than they are. When you assume ascendancy there is an attempt to be moved or guided in a particular direction or along a particular course. To some extent, this could be an extension of a surrogate; (to put into the place of another as a successor, substitute, deputy; substitute for another).

## Article

National Institute of Mental Health, Thomas Insel: A Plan for Changing Times, March 26, 2015

**Context** *Research has transformed many areas of medicine, with profound effects on morbidity and mortality. Exciting advances in neuroscience and genomics have transformed research but have not yet been translated to public health impact in psychiatry. Current treatments are necessary but not sufficient for most patients.*

**Objectives** *To improve outcomes we will need to (1) identify the neural circuitry of mental disorders, (2) detect the earliest manifestations of risk or illness even before cognition or behavior appear abnormal, (3) personalize care based on individual responses, and (4) implement broader use of effective psychosocial interventions.*

**Results** *To address these objectives, NIMH, working with its many stakeholders, developed a strategic plan for research. The plan calls for research that will (1) define the pathophysiology of disorders from genes to behavior, (2) map the trajectory of illness to determine when, where, and how to intervene to preempt disability, (3) develop new interventions based on a personalized approach to the diverse needs and circumstances of people with mental illnesses, and (4) strengthen the public health impact of NIMH-supported research by focusing on dissemination science and disparities in care.*

**Conclusions** *The NIMH is shifting its funding priorities to close the gap between basic biological knowledge and effective mental health care, paving the way for prevention, recovery, and cure.*

1. <u>Successor:</u> The person that comes after another. It may not be a "finish-to- start": the relationship; the goal is to have a command of the present before progress is to continue.

2. Substitute: A person or integrated force that takes the place of it's being with various parts or aspects linked or coordinated.

3. Deputy: A go between handlers. To express or interpret another's views.

4. Substitute: For another; to satisfy the same needs for the individual. It can be used to replace one another. It must share a particular relationship with the good of the individual.

## This or that do I have a choice?

When choosing from an array of impressive and assortment of help; what is a person to look for?
(Acts Ch. 1: verses 20-22).

*What is the best purchase decision for* a person needs?
(Luke Ch. 11: verses 49-50).

Is it difficult to distinguish between two or more things?
*(Malachi Ch. 2: verses 1-5).*

# Secular Crossword Puzzle
## Exercise - Chapter Two

## Across:

2 - _____; Deranged function in an individual or an organ due to a disease.

4 - _____; It applies recombinant DNA, DNA sequencing methods, and bioinformatics to sequence;

7 - NIMH; refers to, National_____of Mental Health.

8 - _____circuitry; A functional entity of interconnected neurons that is able to regulate its own activity using a feedback loop.

10 - _____; Death, especially on a large scale.

## Down:

1 - _____; Take action in order to prevent (an anticipated event) from happening.

3 - _____; A term used to describe a focus on death.

5 - _____; The action or fact of showing an abstract idea.

6 - _____; A person with an interest or concern in something.

9 - _____; It is the scientific study of the nervous system.

# Biblical Crossword Puzzle
## Exercise - Chapter Two

## Across:

3 - And I will curse your_____:

4 - _____; Individuals who are regarded as inspired teachers or proclaimers of the will of God.

6 - Which was_____ from the foundation of the world.

8 - _____; One who is sent out; messengers/ ambassadors of Jesus Christ.

11 - _____; The extent of country over which a bishop has jurisdiction and see.

14 - If ye will not_____,

15 - _____; Deserted of people and in a state of bleak and dismal emptiness.

16 - And spread_____ upon your faces.

17 - _____; Any expressed wish that some form of adversity or misfortune will befall or attach to some other entity.

## Down:

1 - For it is written in the book of_____,

2 - To give_____unto my name,

5 - _____; A person or thing giving or serving as evidence.

7 - _____; Greatly impress or amuse (someone).

9 - _____; Very serious or formal in manner.

10 - O ye priests, this_____ is for you.

12 - _____; A sacred agreement between God and a person or group of people.

13 - That the_____of all the prophets,

# Chapter Three

## Hail. Hell, the gangs all here.

### Article

***The Effectiveness of Assertive Community Treatment for Homeless Populations With Severe Mental Illness: A Meta-Analysis*** Craig M. Coldwell, M.D., M.P.H. William S. Bender, M.P.H.

1. <u>Inclusion</u>; The act of including someone in something. The inherent worth and dignity of all people are recognized.

2. <u>Assertive vs. Standard</u>; Focusing on the goals that are intended; the assertive person uses a conversational tone for dialogue. The message, body expression and eye contact are working together. Values of self and others are important because whether it's one person or a group, the goal is a "win –win" result. Standard; tends to fluctuate depending on situations involved. Despite personal feelings with self or others, they will go along just to get along, interaction will be none, or very little,

oftentimes individual goals or group goals are misplaced and misdirected.

3. <u>Community</u>; A group of people with a common characteristic or interest.

4. <u>Structured</u>; Having a limited number of cored or nearly correct elements.

5. <u>Controlled</u>; to verify a specific experiment by conducting a parallel experiment in which the variable being investigated is held constant or is compared with a standard.

6. <u>Homeless</u>; The condition to acquire and maintain regular, safe, secure and adequate housing.

7. <u>Multidisciplinary Team</u>; A team of professionals including representatives of different disciplines who coordinate the contributions of each profession which are not considered to overlap, in order to improve patient care.

Since no one wants to be around me because of the way I am; I'll just go where I'm accepted. If being displaced gives you power and peace of mind; do be it. Whether it is a hand up, or a hand out; what is your end result? Where are my roots, where is my identity, where is my security, do I belong in this place? So I belong in this setting? House, home, under a bridge, part way in a tunnel, is there really much of a difference? Think of it this way; home is where the mind is.

## When you know God, you
## know who the devil is.

Can I have strength in the face of my pain and grief? Must I use all of my courage to face the ordeal? In my mind and in my body is there a "Tug-of- War" between courageousness and fearlessness? Is it an emotion that's causing a strong uncomfortable response to a perceived provocation, hut or threat?

(James Ch., 3: verses 1- 18).

# Secular Crossword Puzzle
## Exercise - Chapter Three

Across:

14 - _____; An idea or opinion that is shared by all the people in a group.

4 - _____; The quality or state of being diverse in character or content.

6 - _____; To determine or ascertain by mathematical methods.

8 - _____; The ability to produce a desired or intended result.

11 - _____; A method of constructing new data points within the range of a discrete set of known data points.

13 - _____; An event or a group of events occurring as part of a sequence; an incident or period considered in isolation:

Down:

1 - _____; The total number of persons inhabiting a country, city, or any district or area.

2 - _____; A level of quality, achievement, etc., that is considered acceptable behavior.

3 - _____; The characteristics of a work of art in which forms and rhythms are defined chiefly in terms of line.

5 - _____; Causing discomfort or distress by extreme character or condition.

7 - Meta-_____; A quantitative statistical analysis of several separate but similar experiments or studies; pooled data.

9 - _____; Disposed to or characterized by bold or confident statements and behavior.

10 - _____; To get, pull, or draw out.

12 - _____; Unlike, different.

# Biblical Crossword Puzzle
## Exercise - Chapter Three

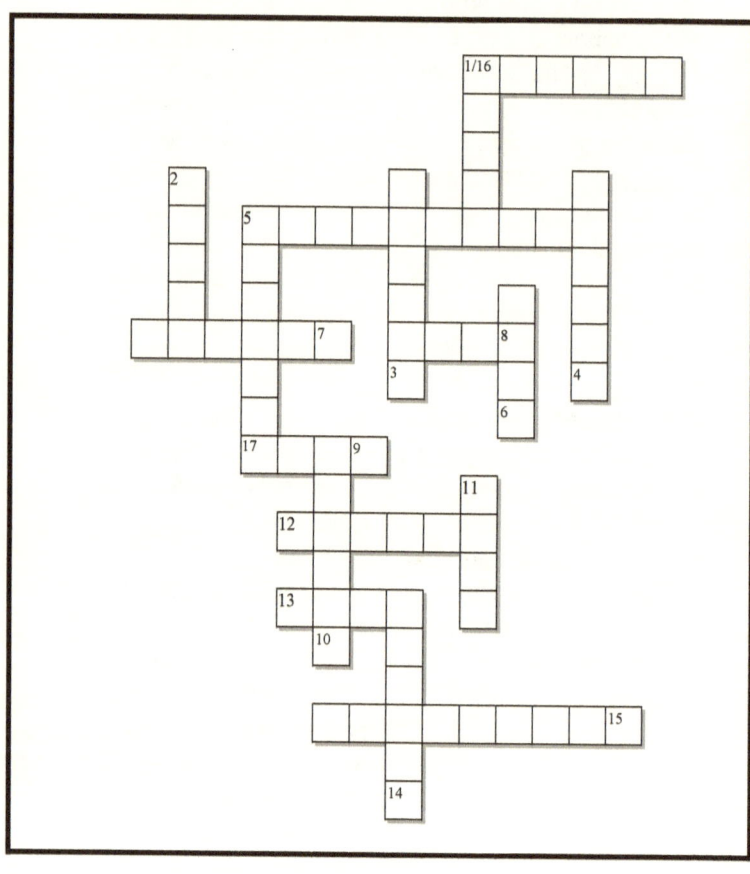

**Across:**

1 - _____; Cause to feel upset, annoyed, or resentful.

5 - _____; A comparison between two things.

7 - _____; Angry, hurt, or resentful because of one's bad experiences or a sense of unjust treatment.

8 - But the tongue can no man _____;

9 - _____; A position of leadership or control.

12 - _____; Very angry or violent disagreement between two or more people or groups.

13 - And the tongue is a _____,

15 - _____; The practice of claiming to have moral standards or beliefs o which one's own behavior does not conform; pretense.

**Down:**

16 - These things _____ not to be.

2 - That it defileth the _____ body,

3 - Full of _____ poison.

4 - _____; An affair or situation under consideration.

17 - Be not many _____,

6 - We put _____ in the horse's mouth,

10 - Even so the tongue is a _____ member,

11 - And it is set on fire of _____.

14 - _____; Show one's resentment or anger, especially by throwing up the head and drawing in the chin.

# Chapter Four

## How do you like me now?

Now, see what you've done; I thought I was okay, until you became involved. Somehow every time you come around me there is a distaste in my mouth and a wound that just won't heal. Hearing your voice and to see your face annoys and exasperate the situation. My blood pressure becomes unstable and my heart rate becomes unstable. What do you want from me?

## Article

"The Over Diagnosis of Depression in Non-depressed Patients in Primary Care" Author, Aragones E. Pinol JL, Labad A

Abstract Depression is one of the most common and debilitating psychiatric disorders and is a leading cause of suicide. Most people who become depressed will have multiple episodes, and some depressions are chronic. Persons with bipolar disorder will also have manic or hypomanic episodes. Given the recurrent nature of the disorder, it is important not just to treat the acute episode, but also to protect against its return and the onset of subsequent episodes.

Several types of interventions have been shown to be efficacious in treating depression. The antidepressant medications are relatively safe and work for many patients, but there is no evidence that they reduce risk of recurrence once their use is terminated. The different medication classes are roughly comparable in efficacy, although some are easier to tolerate than are others. About half of all patients will respond to a given medication, and many of those who do not will respond to some other agent or to a combination of medications. Electro-convulsive therapy is particularly effective for the most severe and resistant depressions, but raises concerns about possible deleterious effects on memory and cognition. It is rarely used until a number of different medications have been tried. Although it is still unclear whether traditional psychodynamic approaches are effective in treating depression, interpersonal psychotherapy (IPT) has fared well in controlled comparisons with medications and other types of psychotherapies. It also appears to have a delayed effect that improves the quality of social relationships and interpersonal skills. It has been shown to reduce acute distress and to prevent relapse and recurrence so long as it is continued or maintained. Treatment combining IPT with medication retains the quick results of pharmacotherapy and the greater interpersonal breadth of IPT, as well as boosting response in patients who are otherwise more difficult to treat.

The main problem is that IPT has only recently entered clinical practice and is not widely available to those in need. Cognitive behavior therapy (CBT) also appears to be efficacious in treating depression, and recent studies suggest that it can work for even severe depressions in the hands of experienced therapists. Not only can CBT relieve acute distress, but it also appears to reduce risk for the return of symptoms as long as it is continued or maintained. Moreover, it appears to have an enduring effect that reduces risk for relapse or recurrence long after treatment is over. Combined treatment with medication and CBT appears to be as efficacious as treatment with medication alone and to retain the enduring effects of CBT. There also are indications that the same strategies used to reduce risk in psychiatric patients following successful treatment can be used to prevent the initial onset of depression in persons at risk. More purely behavioral interventions have been studied less than the cognitive therapies, but have performed well in recent trials and exhibit many of the benefits of cognitive therapy. Mood stabilizers like lithium or the anticonvulsants form the core treatment for bipolar disorder, but there is a growing recognition that the outcomes produced by modern pharmacology are not sufficient. Both IPT and CBT show promise as adjuncts to medication with such patients. The same is true for family-focused therapy, which is designed to reduce interpersonal conflict in the family. Clearly, more needs to be done with respect to treatment of the bipolar disorders. Good medical management of depression can be hard to find, and the empirically supported psychotherapies are still not widely practiced. As a consequence, many patients do not have access to adequate treatment. Moreover, not everyone responds to the existing interventions, and not enough is known about what to do for people who are not helped by treatment. Although great strides have been made over the past few decades, much remains to be done with respect to the treatment of depression and the bipolar disorders.

1. Suicide: Taking an action that is destructive to "yourself" interests. They do not want to continue living; ruin of one's interests or prospects through one's own actions.

2. Recurrent: Turning back as to reverse direction. Good or bad, it keeps coming back.

3. Episode: A separate part of a serialized work; A finite period in which someone is affected by a specified illness.

4. Medications: A drug that can be obtained only by means of a physician's prescription. The art or science of restoring or preserving

health and stable physical, mental and emotional condition by means of drugs.

5.  Psychotherapy (PT): A general term; you learn about your condition and your moods, feelings, thoughts and behaviors.

6.  Cognitive Behavior Therapy (CBT): The goal is to change the person's way of thinking and/or behavior; what are the sources behind your difficulties? A practical approach to problem solving.

7.  Mood Stabilizers: Manic or hypomanic symptoms (acute mania), depressive symptoms (acute depression). The goal is to prevent recurrence. A group of medications that can treat both and rarely cause the other to become worst in the process.

## When it becomes the narrative

The spoken word of God is a clearly stated truth. (Proverbs Ch., 3: verses 1-7).

Leaving no room for confusion or doubt, he has revealed to us and expressed to us the solution. (Isaiah Ch., 55: verses 7-11).

The power of God is free from evasiveness and his will; proceeds in a straight course and manner. (Romans Ch., 12: verses 1-2).

As humans we must allow our extraordinarily and uncomplicated minds to come to an end. (Philippians Ch., 2: verses 5-11).

# Secular Crossword Puzzle
## Exercise - Chapter Four

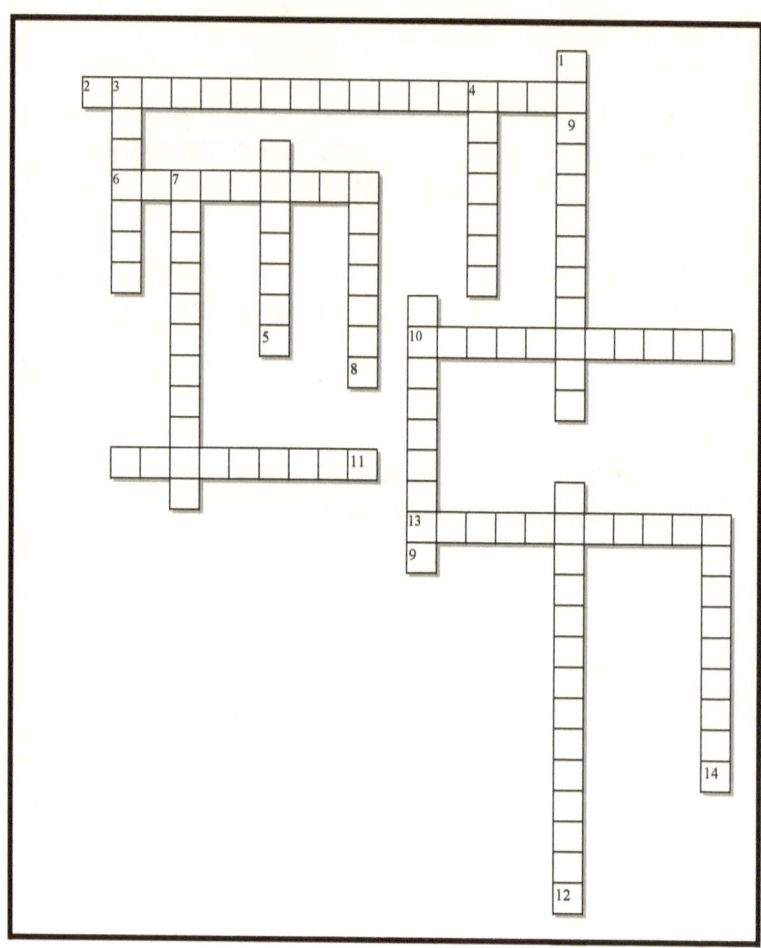

## Across:

2 - _____Therapy; (ECT) is a psychiatric treatment in which seizures are electrically induced in patients to provide relief; psychiatric illnesses.

6 - _____; An emotional state characterized by a distinct period of persistently elevated, expansive, or irritable mood.

10 - _____; Provable or verifiable by experience or experiment.

11 - _____Behavioral Therapy (CBT) is based on the idea that our thoughts cause our feelings and behaviors.

13 - _____; Capable of having the desired result or effect.

## Down:

1 - _____; To make feeble; weaken.

3 - _____; It is used to treat the manic episodes of bipolar disorder.

4 - _____; The act of intentionally causing one's own death.

5 - _____; Also known as manic-depressive illness, is a brain disorder that causes unusual shifts in moods, energy, activity levels.

7 - Initial_____assessment of a person typically begins with a case history and mental status examination,

8 - _____; Having an illness persisting for a long time or constantly recurring.

9 - _____; Someone in a state of general unhappiness or despondency.

12 - _____Psychotherapy; (IPT) is based on the principle that relationships and life events impact mood and that the reverse is also true.

14 - _____; A disturbance or derangement that affects the function of mind or body; example; eating, or the abuse of drugs.

# Biblical Crossword Puzzle
## Exercise - Chapter Four

**Across:**

2 - That ye_____your bodies a living sacrifice,

5 - Fear the lord, and_____from evil.

7 - Being in the_____ of God,

10 - _____; To ask for something in a way that shows you need it very much.

11 - And he shall_____thy paths.

13 - _____; A method, style, or manner of doing something.

15 - _____the wicked forsake his way,

16 - And lean not unto thy_____understanding.

19 - And that every tongue shall_____that Jesus Christ is Lord.

**Down:**

1 - _____; The fact or quality of being alike; resemblance.

3 - _____; The action of forgiving or being forgiven for an error or offense.

4 - Shall they_____to thee.

6 - _____; The ideas or arrangements of ideas that result from thinking.

8 - But be ye transformed by the_____of your mind,

9 - _____; Comply with rules, standards, or laws.

12 - _____; To allow or permit.

14 - And took upon him the_____of a servant.

17 - Let this_____be in you,

18 - And_____him return unto the lord,

# Chapter Five

## Are you out of your mind?

*The word schizophrenia comes from two Greek words that mean to split and mind, because there is a 'split' between what's going on in the person's mind and what is actually happening. There are no medical tests that can be used to say if a person has schizophrenia or not, so getting a diagnosis depends on which list of symptoms are used. It also depends on the doctor or psychologist who talks to the person.*

*Schizophrenia has many different symptoms, and not everyone with schizophrenia has all of them. For this reason, some scientists think that schizophrenia is several separate illnesses that have some of the same symptoms. These scientists claim that the research done on schizophrenia is not accurate since different researchers mean different things when they use the word "schizophrenia" in scientific studies.*

*People with schizophrenia often have delusions or hallucinations. A schizophrenic delusion is a belief that is very different from what other people with the same way of life believe. Hallucinations are usually experiences of hearing voices that don't exist. These voices often say unpleasant things to the person. Many people can hear voices like this without being schizophrenic, for instance right before falling asleep.*

1. <u>Delusions:</u> False beliefs that are beyond your control. Comments seem to be directed at yourself.

2. <u>Hallucinations:</u> All of your "senses" can be affected. You see things and hear things that are extreme.

3. <u>Disorganized Thinking:</u> A person's response and reactions are altered when asked a question, comment, or given a response to a situation. Speech and communication can be impaired.

4. <u>Abnormal Motor Behavior:</u> Unpredictable agitation is the behavior that is exhibited. It is difficult to do tasks and the posture of the individual is inappropriate or bizarre.

5. <u>Negative Symptoms:</u> The person has a monotone behavior and will lose interest in everyday activities. This person lacks eye contact, lacks facial expression, and lacks personal hygiene and the ability to experience pleasure.

Settling in a situation that is far away from nowhere allows a person to explore its reality. If my name is called I pretend to be out; or I am pre-occupied elsewhere. Be unavailable allows one to find use in a new sense of uncharted thoughts. Having an axe to grind, being devoured by rhetoric, and speaking as a means of communication and persuasion.

## Do you want to live? Then keep your mind

The imperfections of humans are consistent and occur daily. The folly that is in us often makes us want to rethink our ways and behaviors. The thought of wanting to better the "self" says, that hope is in reach; and that there is a starting point for improvement.
(2 Peter Ch., 2: verse 4).

When efforts are being made to take a stand, should it not be done on a solid and firm foundation? It is difficult to build on wavering sand.
(Genesis Ch., 3: verse 19), (James Ch., 2: verses 17-18).

# Secular Crossword Puzzle
## Exercise - Chapter Five

**Across:**

2 - _____; Therapy by a professional.

5 - There are over a_____medical problems that can cause the same symptoms of schizophrenia.

8 - _____; False beliefs of danger.

9 - A combination of what has happened to a person and the person's_____may play a role in the development of schizophrenia.

10 - A person with schizophrenia does not change between different personalities: they have only_____.

11 - _____; Not organized.

**Down:**

1 - Some scientists think that schizophrenia is several separate illnesses that have some of the same_____.

3 - _____; It comes two Greek words that means to split and mind.

4 - The symptoms of schizophrenia fall into_____ main categories.

6 - Living in a city during childhood or as an adult has been found to_____the risk of schizophrenia.

7 - _____; is one of the many risk factors that may cause a person to develop schizophrenia.

12 - People with schizophrenia may also have other mental health disorders, like depression, _____and drug abuse.

13 - _____; It is a medical disorder of the mind.

# Biblical Crossword Puzzle
# Exercise - Chapter Five

## Across:

2 - Faith without works_____dead?

4 - Being_____.

21 - _____; Cause light or shadow to appear on a surface.

8 - _____; Fine dry powdery particles, (as of earth).

9 - _____; Activity involving mental, physical or spiritual effort done in order to achieve a purpose or result.

11 - For if God spared not the_____that sinned,

14 - Even so_____,

16 - If it hath not_____,

18 - _____; Used to emphasize something surprising or extreme.

19 - I_____works:

20 - _____dead,

## Down:

1 - _____; In the absence of;

3 - _____; Kept or set apart for some particular use or purpose.

5 - A man may_____,

6 - Thou believest that_____is one God;

7 - The devils_____believe,

9 - _____; Sensible; advisable.

10 - _____; A restrictive force or factor.

12 - _____; A limited or defined extent of the earth's surface, land.

13 - _____; The act of defining; or of making something definite, distinct, or clear.

15 - Shew me thy faith_____thy works,

17 - In the_____of thy face shalt thou eat bread,

# Chapter Six

## Do you want a piece of my mind?

### 1. Situation        2. Conduct        3. Group

### Exercise

In chapters 6, 7 and 8, are set of interrelated tasks to be executed in a group setting. The goal of the situation, conduct and group exercise is to gain knowledge and perception of a situation or fact. The Names, Location, and certain information have been omitted. Each of the following chapters is design to bring understanding to mental illness and promote a greater awareness in relationships and the larger community.

### 1. Situation

C_____ T_____ is a 33 year old woman who is originally from Africa, but is now living in the United States. She is single, has no children, 6 feet 4inches tall and weighs 170 pounds. She has a Bachelor's degree in physical Education. After being unemployed for 31/2 years she now works for a company as a telemarketer. She has recently been diagnosed with "Bipolar depression" and is having difficulty accepting the facts. She refuse to take the prescribed medications and strongly believe that with her diet which consist of fruits, vegetables, and fish, will make her health whole again.

## 2. Conduct

1. She often has feelings of guilt.

2. She has difficulty concentrating.

3. She has noticed a change in her appetite; her weight may go up or down.

4. At times she may have loss of energy.

5. Her interests and enjoyment from things that were once pleasurable are non- existence.

6. Sometimes she feels physically sluggish.

7. Sometimes she feels agitated.

8. Sometimes she has an inability to sleep.

9. Sometimes concentrating becomes difficult

## Questions:

1. Which one of the nine conduct behaviors is the most challenging for this individual? (Explain).

_____

_____

_____

_____

_____

_____

_____

_____

_____

2. Which one of the nine conduct behaviors is the least challenging for this individual? (Explain).

_____

_____

_____

_____

_____

_____

_____

_____

_____

# 3. Group Exercise

Mental Health Professionals:

1. Clinical Psychologist
2. School Psychologist
3. Clinical Social Worker
4. Licensed Professional Counselor
5. Mental Health Counselor
6. Certified Alcohol and Drug abuse Counselor
7. Pastoral Counselor

## Questions

1. Which one of the seven professionals listed, would most effectively address the needs of the individuals? (Explain).

_____

_____

_____

_____

_____

_____

_____

_____

_____

_____

2. Which one of the seven professionals listed would be least effective in addressing the needs of the individual? (Explain).

_____

_____

_____

_____

_____

_____

_____

_____

_____

_____

**When it resonates within self**

Situation

Mark Chapter 5: verses 24-34.

# When receiving the increase, the hold of the decrease had to leave.

1. He has a blood issue;
2. She has a money issue;
3. She has a doctors issue;
4. She has issues with people not accepting her;

5. She has issues with Jewish law, ceremonially unclean health, and attending worship service;
6. She has an issue with humans touching other humans, but she is not included;
7. Twelve long years is an issue for her.

## Conduct

1. Question: If you were this woman with an issue of blood, how would you respond to the people around you? Please choose two phases listed below and explain.

- Don't say nothing and there won't be nothing!
- What the H_____ are you looking at?
- Can I help you?
- Don't let the door hit you on the way out!
- Did not your parents teach you not to stare?

_____
_____
_____
_____
_____
_____
_____
_____
_____
_____

## 3. Group Exercise

Questions: In ten words or less, explain each of the following;

    1.  If I may touch_____
    2.  I shall be whole _____
    3.  Turned him about in the press_____
    4.  Who touched me _____
    5.  Faith_____
    6.  Truth_____

Question: Share and tell us your blessings in your life thus far

_____
_____
_____
_____
_____
_____
_____
_____
_____
_____

# Secular Crossword Puzzle
## Exercise - Chapter Six

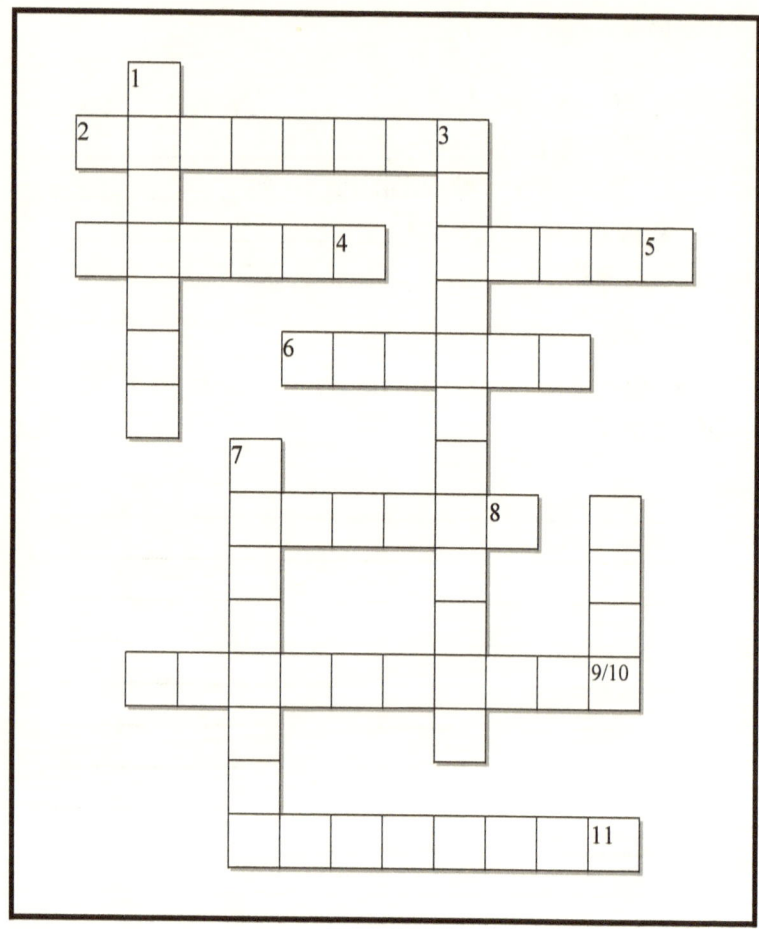

**Across:**

2 - _____Counselor; offers a relationship to that understanding of life and faith; both Psychological and Theological resources.

4 - _____Psychologist; healthy development of children, once called guidance counselors, advocates for students' well-being.

5 – C T is a 33 year old_____.

6 – C T is originally from_____.

8 - _____Health Counselor; cognitive therapy is among the most common technique; often work as part of a health care team.

10 - C T was diagnosed with Bipolar_____.

11 - _____Social Worker; behavioral and bio-psychosocial problems and disorders; 250.000 practitioners, largest group in the nation.

**Down:**

1 - _____Depression; serious shifts in mood, energy, thinking, and behavior; the cycles lasts for days, weeks, or months.

3 - Licensed_____ Counselor; Master's-Degreed (LPCs), trained to work with individuals, families, and groups in treating mental problems.

7 - _____Psychologist; Psychological techniques, cognitive-behavioral therapy, psychoanalytic therapy, they do not prescribe medications.

9 - Certified Alcohol and_____Abuse Counselor; is eligible to be employed and provide services in any state; global standards.

# Biblical Crossword Puzzle
## Exercise - Chapter Six

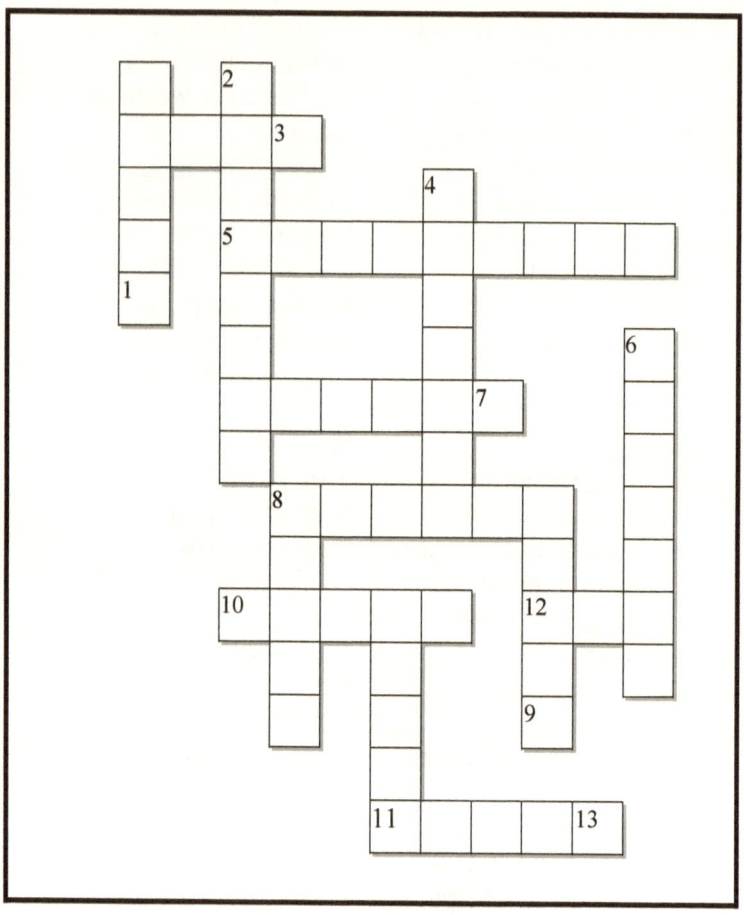

**Across:**

3 - Came and_____down before him,

5 - _____; (of a crowd) fill or be present in (a place or area).

7 - _____; Behavior showing high moral standards.

8 - _____; A highly infectious, usually fatal, epidemic disease;

10 - Thy_____hath made thee whole;

12 - For_____said,

13 - _____; A crowd or crowded condition.

**Down:**

1 - _____; Totality lacking no part, member, or element.

2 - _____; Improve on or surpass (an existing or previous level or achievement).

4 - _____what was done in her,

6 - _____; Emotionally affected. Moved.

8 - Go in_____,

9 - _____; The act of sending out or putting forth.

11 - And had_____all that she had,

# Chapter Seven

## Who says I won't?

Situation

M_____ M_____ is a 27 year old male from Mexico living illegally in the United States. His wife and three children are in Mexico waiting for him to return so they can join him and live a better life in the United States. M_____ M_____ is associated with the following symptoms;

- Insane Automatism
- He finds work wherever possible
- He wants to change his life physically and spiritually
- He speaks very little English
- He lives at various residences
- He has several different aliases
- He is reluctant to talk about his past

## 2. Conduct

1. Insane automatism vs., Non-Insane automatism (Explain)

_____

_____

_____

_____

_____

_____

_____

_____

_____

_____

_____

2. The State of the Mind (Explain)

_____

_____

_____

_____

_____

_____

_____

_____

_____

3. The Nature and Quality of the Act (Explain)

_____

_____

_____

_____

_____

_____

_____

_____

_____

_____

## The Verbal Bubble

The Damascus trip brings new air;

1. Saul of Tarsus, (later called Paul), an evil man causing harm to men and women. (Acts Ch., 7: verses 57-60).

2. While traveling in his bubble he fell to the ground and burst his bubble. (Acts Ch., 9: verses 3-5).

3. Getting a breath of fresh air. (Acts Ch., 9: verse 6).

4. The bubble is dead and new air has consumed him. (Acts Ch., 9: verse 20-21).

### 2. Conduct

Question;

Analyzing the community in which you presently live, choose one of the following terms that best describe your behavior towards them.

• Damnation on them/to them

- Don't let the door hit you on the way out!
- Shame on you
- Kick him where the sun don't shine
- You're more than welcome to stay here if you want?
- If you need anything let me know
- I haven't heard from you in a while, is everything all right?

Explain:

_____
_____
_____
_____
_____
_____
_____
_____
_____
_____
_____

# Group Exercise

Question:

If you are having a problem with a family member how do you resolve the problem? Choose one of the following phrases.

- Ah, he'll/she'll, get over it.
- Is it a moment, or a movement?
- What's good for the goose is good for the gander
- All I know to do is pray

Explain:

_____

_____

_____

_____

_____

_____

_____

_____

_____

_____

_____

# Secular Crossword Puzzle
## Exercise - Chapter Seven

## Across:

2 - _____; A false name used to conceal one's identity.

5 - In legal terms, non-insane automatism is linked to_____factors; example; a blow to the head.

6 - _____; The performance of actions without conscious thought or intention.

8 - In legal terms, insane automatism is caused by a disease of the_____.

10 - M_____M_____ is originally from_____.

## Down:

1 - _____automatism; is when a person is mentally ill and commits the crime.

3 - Non-insane_____; is when the accused was temporarily affected; example; while sleepwalking or suffering from a fit or concussion.

4 - M_____M_____ has_____children.

6 - M_____M_____is known for using several_____.

7 - Presently, M_____M_____ wife is living in_____.

9 - The age of M_____M_____ is twenty_____.

# Biblical Crossword Puzzle
## Exercise - Chapter Seven

**Across:**

3 - Is not this he that destroyed them which called on this name in_____,
4 - And they_____Stephen,
6 - And_____with a loud voice,
10 - _____; To cause mental or emotional discomfort; arouse or provoke to action.
12 - _____art thou,
14 - _____; The capital, city of Syria; Nicknamed as the city of Jasmine.
15 - And straightway he_____Christ in the synagogues,

**Down:**

1 - Whose name was_____.
2 - He came_____Damascus:
5 - And he fell to the_____,
7 - And_____their ears,
8 - _____; To show opposition; resist; to protest.
9 - _____; Throw forcefully in a specified direction.
11 - And it shall be_____ thee what thou must do.
13 - And_____him out of the city,

# Chapter Eight

## Don't throw me away

### 1. Situation

J_____ K_____ is a 26 year old male who lives in the United States. He comes from a single parent home, has three sisters ages, 32, 30 and 27. He lives at home with mother and works as a part time dancer. He has been diagnosed as having "Dissociative Identity Disorder" The following is a list of some of his behaviors;

- As a child he loved playing will dolls
- His sisters would always involve him in their activities
- He refuses to mingle with the male gender
- He loves women clothes and the feel of the material
- His dream is to be a model and open up a clothing store
- He is a musician and enjoys making people smile

### 2. Conduct

J_____ K_____ has four distinct personalities;

1. He was born a male but feels free as a woman,
2. Presently, no one knows that he dances on stage as a woman (with makeup, etc.),
3. He is a male prostitute (dressed as a woman),
4. He is the organist at the local church where he worships.

Question:

What are your views on this individual's "Detachment" (Explain),

_____
_____
_____
_____
_____
_____
_____
_____
_____
_____
_____

Question:

What are your views on this individual's "Compartmentalization" (Explain)

_____
_____

_____

_____

_____

_____

_____

_____

_____

_____

## 3. Group Exercise:

Question:

Which one of the following terms/phrases best describes this individual?

- Fake News Media
- There is a better way
- You do what you have to do to make it
- Life is hard
- You only live once
- Who cares what other people think?
- I'm liven
- Why are you hiding?
- Seek counseling
- You laugh to keep from crying

Explain:

_____

_____

_____

_____

_____

_____

_____

_____

_____

_____

_____

# The Temperament to Succeed

## Genesis Chapter 19: verses 1-38.

### 1. Situation

What does it take for a person to succeed in life and how does a person maintain a sound mind? How does a person overcome personal problems, and prosper in life? God is able to deliver us from the wrath to come.

1. Was Lot saved? 2 Peter Ch., 2: verses 7-8. _____

2. Lot moved to Sodom for a better life. Was this a smart move? Genesis Ch., 13: verse 10. _____

3. Can a person's money and position save them? Ezekiel Ch., 16 verse 49._____

### 2. Conduct

Sodom is an example of a place where God is not present. The city that offers;

- Vain behaviors
- Living to satisfy their lusts
- Homosexuality

- The city was prosperous
- The city was sophisticated
- Behaviors of arrogance
- A city with abundant food supply

Question;

Do you, or do you know someone who need help with any of the issues listed? Explain:

_____
_____
_____
_____
_____
_____
_____
_____
_____
_____
_____

## 3. Group Exercise

Question:

What are your views on homosexuality in our society? Is it a mental illness? Explain:

_____
_____

_____

_____

_____

_____

_____

_____

_____

_____

_____

Question:

If God's word is perfect, why do some people continue to disobey him? Explain:

_____

_____

_____

_____

_____

_____

_____

_____

_____

_____

# Secular Crossword Puzzle
## Exercise - Chapter Eight

**Across:**

2 - _____; Offer (someone) for sexual activity in exchange for payment.

3 - The signs and symptom of dissociative_____ disorder vary depending on the individual.

5 - _____; Physical or verbal representation or simplified version of a concept, relationship, or aspect of the real world.

6 - _____; A small model of a human figure, often one of a baby or girl, used as a child's toy.

8 - Dissociative identity_____; recently defined; diagnostic and statistical manual of mental disorders, 5th Edition.

**Down:**

1 - _____; "Solation" or splitting off or part of the personality or mind with lack of communication and consistency between the parts.

4 - _____identity disorder (DID).

7 - _____; A sense of separation from normal association or environment.

# Biblical Crossword Puzzle
## Exercise - Chapter Eight

**Across:**

4 - And_____just Lot,
5 - Two_____came to Sodom in the evening.
8 - The men of the city surrounded Lot's_____.
11 - But his wife from behind him, looked back, and she became a pillar of_____.
12 - Fullness of_____,
14 - The Lord rained on Sodom and Gomorrah brimstone and_____.
15 - I have two daughters who have not had relations with man; please let me bring them out to_____.
16 - Before the Lord destroyed Sodom and_____,

**Down:**

1 - And beheld all the plain of_____,
2 - And_____lifted up his eyes,
3 - Where are the men that came to you? Bring them out so we can have_____with them.
6 - Behold, this was the iniquity of thy sister Sodom,_____,
7 - Do_____to these men.
9 - Lot was sitting at the_____of Sodom.
10 - _____; Difficult and much debated.
13 - Lot went up from_____, and stayed in the mountains; his two daughters with him; in a cave.

# ANSWERS TO THE PUZZLES

## Chapter 1 Answers To Secular Crossword Puzzle

Across (forward)
6. Supernatural
10. Baghdad

Down (downward)
1. Madhouses
3. Measuring
4. Trephining

Across (backward)
8. Evil
9. Hysteria
11. Asylum

Up (upward)
2. Devices
5. Black
7. Mental

## Chapter 1 Answers To Biblical Crossword Puzzle

Across (forward)
3. Beginning
8. Was
10. First
12. There

Down (downward)
7. Light
9. Spirit

Across (backward)
2. Omega
5. All
6. Said

Up (upward)
1. Divided
4. Face
11. Hell
13. Was

# Chapter 2 Answers To Secular Crossword Puzzle

Across (forward)
2. Pathophysiology
4. Genomics

Down (downward)
5. Manifestations
9. Neuroscience

Across (backward)
7. Institute
8. Neural
10. Mortality

Up (upward)
1. Preempt
3. Morbidity
6. Stakeholders

# Chapter 2 Answers To Biblical Crossword Puzzle

Across (forward)
3. Blessings
4. Prophets
6. Shed
8. Apostles
15. Desolate

Down (downward)
2. Glory
9. Solemn
10. Commandment
12. Covenant
13. Blood

Across (backward)
11. Bishoprick
14. Hear
16. Dung
17. Curse

Up (upward)
1. Psalms
5. Witness
7. Slay

# Chapter 3 Answers To Secular Crossword Puzzle

**Across (forward)**
4. Heterogeneity
6. Calculated
11. Interpolation

**Down (downward)**
1. Subpopulation
2. Standard
3. Linearly
10. Extract

**Across (backward)**
8. Efficacy
13. Episode
14. Consensus

**Up (upward)**
5. Severe
7. Analysis
9. Assertive
12. Dissimilar

# Chapter 3 Answers To Biblical Crossword Puzzle

**Across (forward)**
1. Offend
5. Similitude
12. Strife
13. Fire

**Down (downward)**
2. Whole
11. Hell
16. Ought

**Across (backward)**
7. Bitter
8. Tame
9. Helm
15. Hypocrisy

**Up (upward)**
3. Deadly
4. Matter
6. Bits
10. Little
14. Bridle
17. Masters

# Chapter 4 Answers To Secular Crossword Puzzle

Across (forward)
2. Electroconvulsive
6. Hypomanic
10. Empirically
13. Efficacious

Down (downward)
1. Debilitating
3. Lithium
4. Suicide
7. Psychiatric

Across (backward)
11. Cognitive

Up (upward)
5. Bipolar
8. Chronic
9. Depressed
12. Interpersonal
14. Disorders

# Chapter 4 Answers To Biblical Crossword Puzzle

Across (forward)
5. Depart
10. Beseech
11. Direct
15. Let
16. Own

Down (downward)
1. Likeness
3. Pardon
6. Thoughts
8. Renewing
14. Form
18. Let

Across (backward)
2. Present
7. Form
13. Ways

Up (upward)
4. Add
9. Conformed
12. Let
17. Mind

# Chapter 5 Answers To Secular Crossword Puzzle

**Across (forward)**
2. Psychotherapy
11. Disorganized

**Down (downward)**
1. Symptoms
3. Schizophrenia
4. Three
7. Trauma
12. Anxiety
13. Schizophrenia

**Across (backward)**
5. Hundred
8. Paranoia
9. Genes
10. One

**Up (upward)**
6. Double

# Chapter 5 Answers To Biblical Crossword Puzzle

**Across (forward)**
2. Is
11. Angels
14. Faith
18. Even
19. Have
20. Is
21. Works

**Down (downward)**
1. Without
3. Reserved
9. Well
12. Ground
13. Is

**Across (backward)**
4. Alone

**Up (upward)**
5. Say

6. Say

7. Also

8. Dust

10. Chains

16. Works

15. Without

17. Sweat

# Chapter 6 Answers To Secular Crossword Puzzle

**Across (forward)**
2. Pastoral
6. Africa

**Down (downward)**
7. Clinical

**Across (backward)**
4. School
5. Woman
8. Mental
10. Depression
11. Clinical

**Up (upward)**
1. Bipolar
3. Professional
9. Drug

# Chapter 6 Answers To Biblical Crossword Puzzle

**Across (forward)**
5. Thronging
8. Plague
10. Faith
12. She

**Down (downward)**
2. Bettered
4. Knowing
6. Touched

**Across (backward)**
3. Fell
7. Virtue
13. Press

**Up (upward)**
1. Whole
9. Issue
11. Spent

# Chapter 7 Answers To Secular Crossword Puzzle

**Across (forward)**
5. External
6. Automatism

**Across (backward)**
2. Alias
8. Mind
10. Mexico

**Down (downward)**
3. Automatism
7. Mexico

**Up (upward)**
1. Insane
4. Three
9. Seven

# Chapter 7 Answers To Biblical Crossword Puzzle

**Across (forward)**
3. Jerusalem
6. Cried
10. Pricks
12. Who
14. Damascus
15. Preached

**Across (backward)**
4. Stoned

**Down (downward)**
5. Earth
9. Cast
11. Told

**Up (upward)**
1. Saul
2. Near
7. Stopped
8. Kick
13. Cast

# Chapter 8 Answers To Secular Crossword Puzzle

Across (forward)
2. Prostitue

Down (downward)
1. Compartmentalization
4. Dissociative
7. Detachment

Across (backward)
3. Identity
5. Model
6. Dolls
8. Disorder

Up (upward)
No answers

# Chapter 8 Answers To Biblical Crossword Puzzle

Across (forward)
4. Delivered
5. Angels
8. House
16. Gomorrah

Down (downward)
3. Relations
7. Nothing
3. Zoar

Across (backward)
11. Salt
12. Bread
14. Fire
15. You

Up (upward)
1. Jordan
2. Lot
6. Pride
9. Gate
10. Vexed

# References

## Chapter One

**The History of Mental Illness: From Skull Drills to Happy Pills.**
Allison M. Foerschner 2010, Vol. 2 No. 09 | pg.1-2

Bible KJV
Genesis Chapter 1: verses 1-5.
Psalm 139: verse 8.
John Chapter 1: verses 1-4.
John Chapter 4: verse 24.
Revelation Chapter 22: verse 13.

## Chapter Two

**Translating Scientific Opportunity into Public Health Impact A Strategic Plan for Research on Mental Illness** Thomas R. Insel, MD, *Arch Gen Psychiatry. 2009; 66(2):128-133*

**National Institute of Mental Health**, Thomas Insel: A Plan for Changing Times, March 26, 2015

Bible KJV
Acts Chapter 1: verses 20-22.
Luke Chapter 11: verses 49-50.
Malachi Chapter 2: verses 1-5.

## Chapter Three

**The Effectiveness of Assertive Community Treatment for Homeless Populations With Severe Mental Illness: A Meta-Analysis**, Craig M. Coldwell, M.D., M.P.H. William S. Bender, M.P.H. THE AMERICAN JOURNAL OF PSYCHIATRY March 2007 Volume 164 Number 3

Bible KJV
James Chapter 3: verses 1-18.

## Chapter Four

**Treatment and Prevention of Depression**
Steven D. Hollon,, Michael E. Thase,, John C. Markowitz
Psychological Science in the Public Interest, Volume: 3 issue: 2, page(s): 39-77 Issue published: November 1, 2002

**The Over Diagnosis of Depression in Non-depressed Patients in Primary Care.** Author, Aragones E. Pinol JL, Labad A. Archives of General Psychiatry. 1996; 53:842-848 (PubMed)

Bible KJV
Proverbs Chapter 3: verses 1-7.
Isaiah Chapter 55: verses 7-11.
Roman Chapter 12: verses 1-2.
Philippians Chapter 2: verses 5-11.

## Chapter Five

**Schizophrenia**, https://simple.wikipedia.org/wiki/Schizophrenia

**National Institute of Mental Health** http://www.nimh.nih.gov/health/topics/schizophrenia/index
July, 2016

Bible KJV
Genesis Chapter 3: verse 19.
James Chapter 2: verses 17-20.
2 Peter Chapter 2: verse 4.

## Chapter Six

Bible KJV
Mark Chapter 5: verses 24-34.

## Chapter Seven

Bible KJV
Acts Chapter 7: verses 57-60.
Acts Chapter 9: verses 3-6; 20-21.

## Chapter Eight

**The Diagnostic and Statistical Manual of Mental Disorders, 5th Edition.** Cardwell C. Nuckols, PhD

Bible KJV

Genesis Chapter 13: verse 10.
Genesis Chapter 19: verses 1-38.
Ezekiel Chapter 16: verse 49.
2 Peter Chapter 2: verses 7-8.

# Conclusion

Being without the savageness or fear of humans allows one to have a useful and adjective amount of communication. Having the need to belong and to be accepted by others is often an ongoing task. Enhancing a cause or to feel comfortable in any situation one must have at least a glimpse of hope. Mental illness has embarked upon our families, communities, society, and the world. If mental illness is here to stay, then we as a modern people must accept it.

www.ingramcontent.com/pod-product-compliance
Lightning Source LLC
Chambersburg PA
CBHW030910180526
45163CB00004B/1774